W9-CPD-927

CAREERS INSIDE THE WORLD OF

Technology

You probably don't realize that you use technology every single day.

CAREERS & OPPORTUNITIES

CAREERS INSIDE THE WORLD OF
Technology

by Jean W. Spencer

THE ROSEN PUBLISHING GROUP, INC.
NEW YORK

Published in 1995 by The Rosen Publishing Group, Inc.
29 East 21st Street, New York, NY 10010

First Edition

Manufactured in the United States of America

Library of Congress Cataloging-in-Publication Data

Spencer, Jean W.
    Careers inside the world of technology / by Jean W. Spencer. —
1st ed.
      p. cm. — (Careers & opportunities)
    Includes bibliographical references and index.
    ISBN 0-8239-1896-3
    1. Technology—Vocational guidance—Juvenile literature.
[1. Technology—Vocational guidance. 2. Vocational guidance.
3. Occupations.] I. Title. II. Series.
T65.3.S64   1995
602.3—dc20   95-26955
                                          CIP
                                          AC

Contents

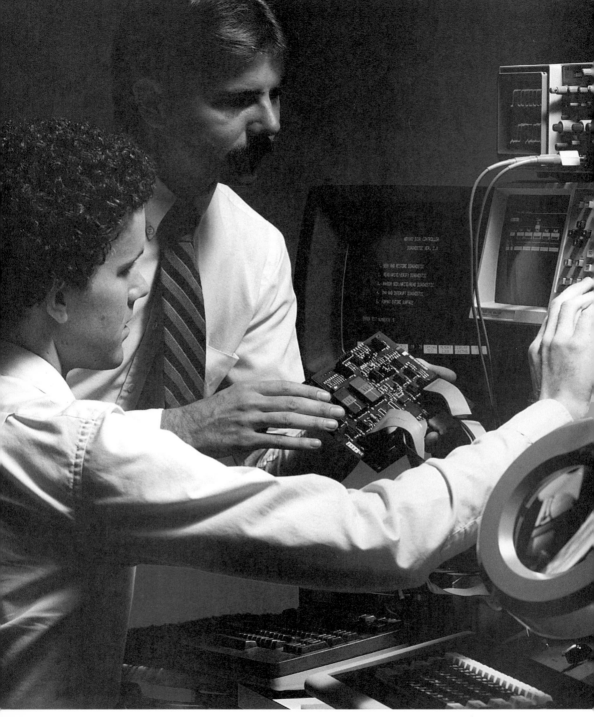

You can put your skills and interests to work as a computer-repair
technician.

Chapter 1

EXPLORING TECHNOLOGY

"After graduation, if you could do anything you wanted to do, what would you choose?"

Ed looked at the question for a long time. It was a group career meeting, and it was his turn to answer. Lou wanted to be a veterinarian. Lawyer was Nancy's choice. Brian said, "Play basketball or teach." Chris talked about show business and big money.

Ed didn't want to sound boring, but he enjoyed fixing things like computers or radios or other things that were broken. "Fix computers," he finally responded.

Ed didn't know it then, but that was one of the best answers he could have given. His answer meant that he liked to *do* things, not just plan, think, or talk about them. It meant that his interest was in the world of technology, a job field that is growing fast.

Technology means using science to solve problems or improve products that people use every day. Computers are one result of technology. Another is the technology required to manufacture things such as clothes, CDs, and cars. A third example includes the machines in the field of health care: x-rays, CAT scans, sonographs, heart monitors, even the computer programs that do the accounting required for the hospital to bill you. All are examples of technology.

Jobs in technology are growing in number largely because computers are revolutionizing many jobs. The adventure of technology is open to those who really want it. Following is a list of some ways you might become part of it.

SOME TECHNOLOGY CAREERS

aerospace technician
aeronautics technician
agriculture
   soil analysis
   hydroponic growing
   watering systems
automotive technology
   emissions control
air conditioning and
    refrigeration
aircraft technician

air traffic control
agribusiness technician
agricultural equipment
   technician
animal health technician
animal production
   technician
appliance repairer
architectural technician
audio technician
audio-control technician

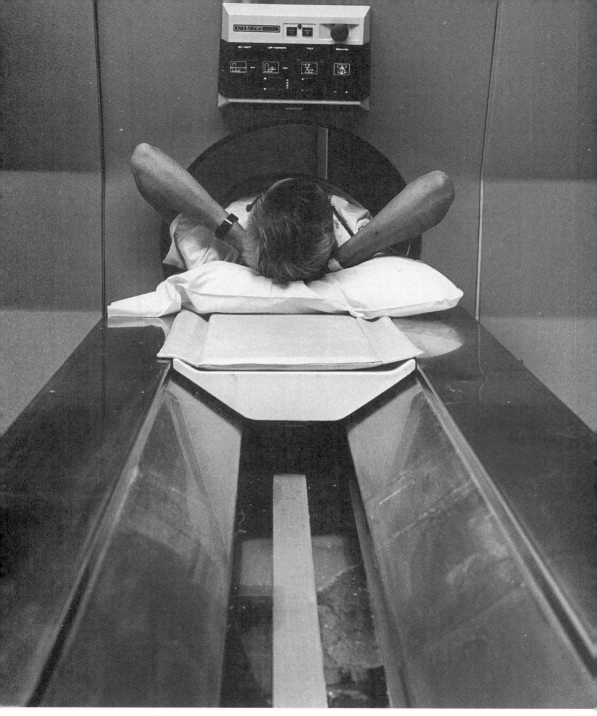
Thanks to technology, important medical diagnostic devices such as the CAT scan have been created.

audiovisual technician
automatic equipment
    technician
automotive technician
automotive engine
    technician
avionics technician
bench work
biomedical technician
biomedical equipment
    technician
broadcasting
cable TV technician
cartography
chemical technician
civil engineering technician
coal mining technician
computer service technician
computer technician
communications technician
computerized design
corrective therapy assistant
crime lab technician
cryptographic technician
darkroom technician
delivery service
dietetic technician
dental laboratory
    technician
dialysis technician
drafting and design
    technician
design technician
diving technician

drafting technician
ECG technician
ecological technician
EEG technician
EKG technician
electrical technician
electrical power technician
electromechanical
    technician
electronic publishing
electronic technician
electronic test technician
emergency medical
    technician
engineering technician
environmental technology
exhaust-emissions
    technician
farm crop production
    technician
fiber optics
field technician
fire safety technician
fire technology
fish production technican
food planning and controls
food technician
forestry and conservation
graphic arts technician
hazardous materials
    control
heating and cooling
home electronics repairer
hot-cell technician

hotel and restaurant
industrial mechanic
industrial technician
industrial electronic
   equipment
information superhighways
installations
instrument repairer/maker
instrumentation technician
integrated circuit board
   technician
laser technician
laser test technician
lights technician
machinist
manufacturing technician
mechanical technician
medical assistant
medical technician
medical lab technician
mobile computing
machine trades
manufacturing technician
manufacturing technologist
mechanical technician
medical electronics
   technician
medical technologist
metallurgical technician
moldmaking
multimedia specialist
nuclear medical
   technologist
nuclear reactor technician

oceanography technician
operations technician
optics technician
orthotic and prosthetic
   technician
perfusionist
pest control
petroleum technician
pharmaceutical technician
photo technician
photographic equipment
   technician
physical therapy assistant
physical radiological
   technician
plastics technician
pollution control technician
power plant worker
nuclear reactor operator
   technician
printed circuit board
   technician
printing-publishing
processing operations
public utilities
quality control
radio technician
repairer
respiratory therapist
robotics technician
science technician
scientific and business data
   processing technician
security systems

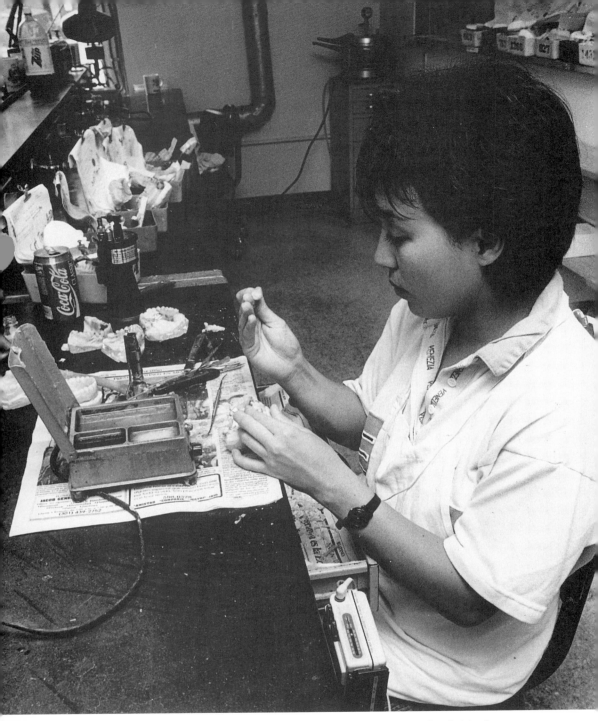

Dental technician is one of the many job opportunities in the field of technology.

semiconductor technician
soft tool technician
soil conservation technician
sound-recording technician
sound effects
sound technician
studio technician
transmitter technician
stage production technician
stage technician
surgical technician
surveying and mapping
  technician
technical writer
telecommunications
  technician
telephone equipment

testing equipment
textile technician
quality control
tool and die maker
TV control room
typesetter
underwater technician
video technician
water and wastewater
  treatment
welding technician
winemaker
wood science and
  technician careers
wood processor operator
word processing
x-ray technicians

Questions to Ask Yourself

The world of technology is always expanding. There are thousands of opportunities for those who want to be in on the future. 1) Take a look at the list of professions. Which appeals to you? Why?

This woman is doing electronics testing.

CHAPTER 2

COMPUTERS AND ELECTRONICS

"**C**omputers can do anything!" Jeff's spare time takes him straight to his home computer. Today he's flying an airplane and then running a city. Yesterday, he sent and received e-mail with friends in five states. Tomorrow, he may download a new software program that will show him how to test his own computer in case of trouble.

Jeff has watched Grampa Jack troubleshoot computers since he was in kindergarten, so he knows about boards and other technological aspects of the inside. He can take his knowledge and love of computers and apply them to a future career—computer technician.

Computer Technicians

Computer service technicians can choose to work for computer manufacturers, large business customers, computer or software sales firms, or private users working out of home offices.

15

Most large companies and governments have computer technicians on staff to save valuable employee time when computers are down or when new software or hardware is being installed. Service is needed quickly to provide an instant fix; they can't afford to wait.

Other computer techs work for small service companies. They may drive to their customers, or operate a sales and repair business to which owners bring their computers for service.

In any location, when a computer problem needs attention, technicians must first find out where the failure is. Next they figure out how to fix it. This process is called troubleshooting.

A first step is to determine whether there are user problems, often called "cockpit" problems. If so, the technician must be able to teach the computer users without embarrassing them.

Electronics Technicians

Electronics labs hire many specialists who make, build, repair, or test equipment. They work at a bench and operate a variety of testing machines.

Paul installs complete computer systems for hospitals or individuals by unpacking each piece of equipment, connecting cables from computers to printers, and setting up software, modems, and e-mail links. Modems provide a way for computer users to transmit information from one computer to another. E-mail is an international computer mailbox through which people can give and receive all kinds of information. Paul is an expert at starting up such systems. His customers depend on him. His business is growing. So is his working knowledge of other electronics gear such as printers, fax-modems, accelerator boards, scanners, and electronic music keyboards.

Integrated Circuit Board Technicians

Give John a schematic and he can build it into a functioning integrated circuit board with transistors, resistors, capacitors, and solder. The integrated circuit board is the basic building block in any kind of electronic equipment that enables it to work. John winds coils and transformers out of copper wire, follows color-coded wiring systems that allow various components to "talk" to each other, and solders the right parts together.

Most of this work is small enough that John frequently uses a magnifier to see the tiny connections clearly. These boards seem mysterious, but to John they are like a puzzle to be solved. He enjoys bringing components together, plugging in the finished product, and watching it work.

Robotics

Robotics technology is growing fast.

Robots are found in manufacturing plants,

among other places. They do not look like "Star Wars" favorites R2D2 or C3PO. Usually they are headless and look like metal arms with movable pincers picking up parts and putting them in place. Many robots swivel in ways that human shoulders and elbows cannot. Robots also make workplaces safe because they protect humans from chemicals, odors, and risky lifts and movements. To see many robotic machines working together is both fascinating and grotesque. Robots at work on an assembly line have an otherworldly effect.

The advantage is that industrial robots do not get bored doing the same thing over and over again as human beings do. The disadvantage is that there are fewer assembly jobs for entry-level job seekers. The good news for you is that there will be new, more interesting jobs for those who learn how to make, install, adjust, program, service, and repair robotics in the workplace.

Questions to Ask Yourself

Computers and electronics are everywhere. There is always demand for someone skilled in these areas. 1) Which profession discussed in this chapter would you prefer? 2) Do you pay attention to details? 3) Could you keep up with the changes in the electronics and computer worlds?

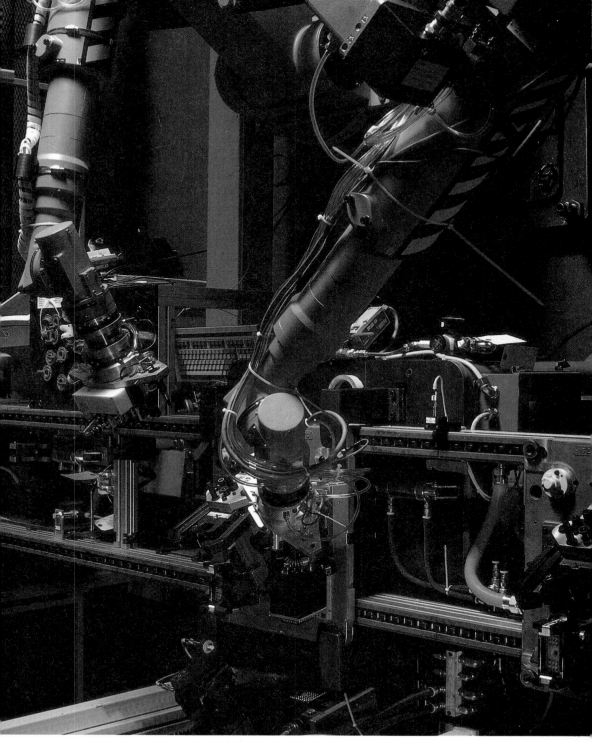

An automated assembly line using robots.

CHAPTER 3

MANUFACTURING TECHNOLOGY

Manufacturing is the process of taking raw materials such as metal, cloth, wood, glass, chemicals, or plastics and making them into something that people want or need. When you make one item it is craftsmanship. When you make thousand or millions it is manufacturing.

Manufacturing is teamwork. When your favorite singer lays down the track for a new song, audio recording technicians work with her to get the sound recorded just right. But if it is going to sell a million, first a million copies have to be made. Manufacturing CDs is a very complex technological process. In this job, you work in a dust-free clean room and wear white coveralls, shoe covers, headgear, and a mask because the tiniest speck of dust can ruin the disc.

Whatever your special interest, you have great opportunity to make a job match in the world of manufacturing technology. People and high-tech machines manufacture 60,000 pairs of Levi's every day. Others manufacture lipsticks and face creams. Still others build racing bicycles.

New technology uses science to make each product better, available, and affordable. At General Motors high tech means speed, performance, and appearance. A GM race driver hit over 250 miles per hour testing Delta's quad-four engine. The whole technology team celebrated the results of their teamwork.

Manufacturing engineers and technologists like problem-solving. Tinkering is useful to them, getting ready for a career and in their work. "Tinkering is the hands-on way to get in there and see things happen," explains Rusty. "It's like a puzzle, getting everything to fit."

Manufacturing technology brings together the best of mechanical, chemical, electrical, and electronic disciplines, blending them to create the best product. The machinery in full operation can be noisy. Workers often wear protective gear, which may include goggles, steel-tipped shoes, or fire-resistant coveralls.

Today's manufacturing plants and mills are being radically transformed by product competition, computerized machines, robotics, and new technology pushed to its limits. A good example is papermaking. Workers are aided by computerized equipment. The process still takes muscle,

It takes the work of both muscle and technology to make paper.

but computerizing the entire operation at a Willamette Industries' mill is changing job requirements and job skills. It is a high-tech operation for the crews.

"The Machine" controls the process 24 hours a day as wood chips and recycled paper products are beaten into pulp, converted into paper or cardboard, dried, wound into rolls, labeled and banded for shipping. This is called automation.

High-tech calls for more training. Increased computer skills are needed for advancement as technology continues to improve manufacturing workplaces.

Environmental Technology

Willamette papermakers grow their own trees to manufacture lumber and plywood. Leftover fiber and chips are used to make paper, particleboard, and fiberboard. Wood and paper products are recycled. Trees are replanted in their forests.

Diving technicians may work on underwater research projects, photography, or offshore oil wells. They may search for treasure long buried on the ocean floor where ships sank 100 or more years ago. Dressed in scuba gear and diving suits, they use oxygen tanks and specialized tools below the water to search for missing equipment or to repair holes in ships.

Science technicians study mysteries of the earth such as earthquakes, tornadoes, hurricanes,

As a diving technician, you have the opportunity to see a world that most people don't even realize exists.

floods, icestorms, and snowstorms. Using computers, they measure living mountains that grow
in the midst of earthquakes, chart oceans so that
ships will know exactly where they are, figure out
the fastest routes to jet around the world, or
count wild animals. Technologists also measure,
test, and study the earth to improve predictions
of earthquakes and other natural disasters that
threaten the safety of people and cities.

Other technicians who are concerned with the
environment work in agribusiness, soil conservation, and farming. Technology is changing farms
into big business. With scientific processes and
equipment, crop technicians study what types of
vegetables, fruits, grains, or flowers to grow.
They research new seeds and use instruments
to analyze soil and water and test nutrients,
fertilizers, and disease and pest-prevention sprays.

It takes complex and expensive machinery
to produce our food. Such machinery includes
computer-controlled animal feeding systems,
automated milking machines, and machinery
for preparing and watering the soil, and planting
and harvesting vegetables and grains. Other technicians work with production of animals and
poultry, improving the feed that will keep them
healthy, prevent disease, and grow them quickly.
All of this equipment and more requires technicians to test, operate, sell, and service.

Water and wastewater treatment plants are

Technicians help garbage disposal plants meet the strict air pollution
control requirements by installing new equipment.

vital for all land use. Harmful chemicals and waste must be removed from water to make it usable. Most of us take safe water for granted. Technical workers cannot. It is technology that everyone depends on to be able to drink the water or grow safe food crops with it. These jobs are vital to survival.

Other technicians choose air technology. Keeping planes flying safely depends on technology teams who know the engines and equipment inside and out. From air traffic control to aviation maintenance, technologists rely on state-of-the-art computerized monitors and equipment.

All of these are environmental technology frontiers and hold the promise of interesting and necessary work.

Questions to Ask Yourself

A career in manufacturing technology would give you the chance to create something necessary for society. 1) Are you interested in manufacturing technology or environmental technology? Which appeals to you more? 2) Do you take pride in your work?

One area of technology is auto manufacturing.

CHAPTER 4

HEALTH CARE TECHNOLOGY

Hospitals require more than just doctors and nurses. Hospitals are literally abuzz with electronics.

Pam discovered that when her technician father first researched and developed blood-flow monitoring equipment. She learned even more about health-care technology when her dad needed heart bypass surgery and had to depend on heart-lung machine technology to live. She chose a technical job in a hospital business office and now operates computers to track costs.

Pam's sister and friends chose other types of medical technology careers. Her sister earned AS (Associate in Science) degrees from two community colleges. She worked first as a respiratory therapist (helping people to breathe) and later as an animal health technician (testing, assisting with surgery, and more). Both require hands-on use of monitoring devices and specialized tech-

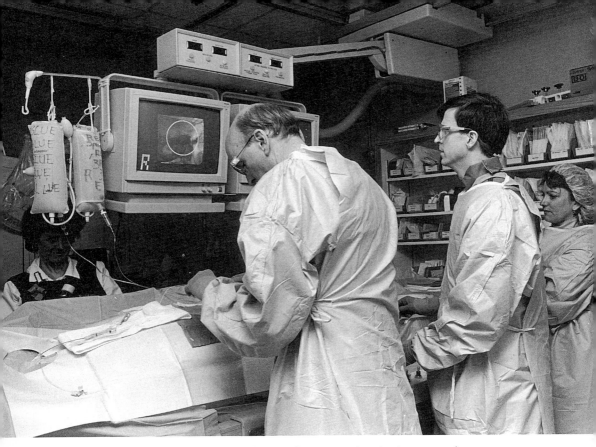

A doctor needs the help of radiation therapy technologists to perform a critical procedure.

nology. Other friends work with word processing technology, accounting systems, data bases, lab technology, and cardiac nursing.

Cardiac-monitor and EKG (electrocardiograph) technicians study heart function and test hearts. Cardiac-monitor technicians work in intensive-care units or cardiac-care units. A cardiac monitor shows how a person's heart is beating, and the technician watches each person's monitor very closely. These specialized monitors look like small video screens with wires attached to the patient's chest to record every heartbeat.

Lead Orthotic Technician Maurice Phillips has been designing and building specialized wheelchairs for patients for over fifteen years.

The technician must be alert for any sign that the heart is not working as it should. If trouble occurs, the technician calls a doctor or nurse for immediate care. This work requires good

31

training, great accuracy, and a strong sense of responsibility.

EEG (electroencephalograph) technicians measure a person's brain waves. Doctors study the results to discover what is wrong and what is needed to help the person be well.

Emergency medical technicians use mobile technology while taking patients to the hospital in ambulances and emergency vehicles.

Amazing medical software programs continually transform computers into machines that perform complex duties, each designed for specific purposes.

Medical equipment operates heart and lungs during surgery, tests blood samples, and performs CAT scans and kidney dialysis. Lasers are used for eye surgery and arthroscopic surgery for sports injuries. All require skilled technicians to operate as well as produce.

X-ray Technologists

Some x-ray technologists specialize in diagnostic medicine, making x-ray photo images with special short light rays to find out exactly what type of disease the patient has, and the exact location in the body. Some of the x-rays are used to locate broken bones or pinpoint cancer growths.

Other x-ray technologists use radiation to remove diseased body tissue very carefully, preserving healthy tissue. Others may concentrate on nuclear medicine, in which radioactive substances are injected or swallowed by the patient to treat particular problems. Still others may operate sonography equipment, which uses sound waves instead of light waves to make an image available for doctors and radiologists to study.

Whatever specialty x-ray technologists are working with, they must be extremely alert and very skilled in using instruments and substances that could be highly dangerous. They must take great pains to offer both comfort and safety to every patient in their care.

Respiratory Therapists

Lu-Anne chose respiratory therapy as her specialty. That means treating patients who have breathing problems. Another name for this specialty is inhalation therapy, because it helps people breathe in (inhale) oxygen. We cannot live without oxygen for more than a few minutes.

Most of us take breathing for granted, but some illnesses such as asthma, emphysema, or polio interfere with normal breathing. Accident, emergency, injury, or some surgeries can also interfere with this process. Then Lu-Anne uses a variety of machines such as respirators or ventilators to restore and strengthen the patient's ability to breathe. As part of a Code Blue team that responds when someone is in cardiac arrest (stopped heartbeat), Lu-Anne can be called to administer oxygen at a moment's notice. She finds it gratifying to be able to help not only the patient, but the families who are upset.

These are only some of a great many paraprofessional medical careers that depend on technology to help patients. They are not easy, but they are rewarding because they help people get well.

Linking Medicine and Business Office
Technology makes a difference in every phase of health care. Incoming patients are listed in a hospital's database as soon as they are admitted. All data remains in the mainframe computer so that it may be updated or referred to as needed. In this way, accurate information is available to authorized staff for every phase of treatment, including billing and insurance services. Making the patient happy and well is the goal of the entire health care staff.

Questions to Ask Yourself
Health care technology provides a unique oppor-tunity to be involved in both the health care world and technology. 1) Would you want to work in a stressful atmosphere such as a hospital? 2) What aspect of health care technology appeals to you most? Why? 3) Would you prefer hands-on work, or something more behind the scenes, such as database entry?

CHAPTER 5

TECHNICIAN SERVICE JOBS

When the computer system goes down right in the middle of college registration, some one has to fix it. If refrigeration fails in a restaurant or the air conditioning shuts down without warning in a crowded movie theater, the problem must be located and fixed swiftly. If a computerized thermostat is sending "cold" messages to the heater in the middle of an icestorm and an office full of people are shivering, some one will become really popular by getting things warm again. Service technicians build careers on coming to the rescue.

Public and Private Service Projects

Would you like to plan and design freeways, airports, bridges, tunnels, overpasses, and other public works? Civil engineering technicians help engineers with these projects. They often work

35

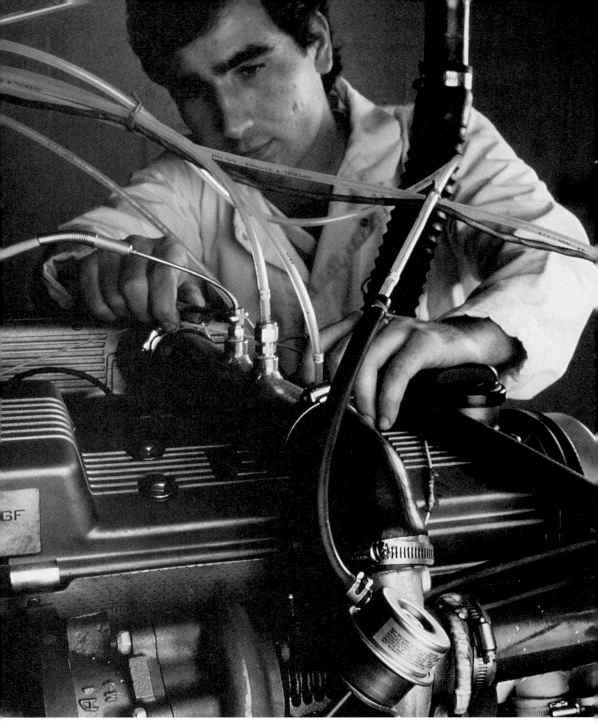

Teams of technicians check out all parts of an automobile before it is released for sale.

outdoors. They need technology skills and organizational skills.

Drafting and design technicians, architectural technicians, and cartographers (mapmakers) work both outdoors and inside. They use computers to make their drawings, measurements, and blueprints faster than ever before.

Automotive Service

When you know what is under the hood, where the power comes from, you have better control of your wheels.

When you decide to drive and turn the key, the electronic system starts the engine and many related parts. Every time you push a button or flip a switch, your signal goes into the car's computer and tells it what to do. If nothing happens, you know the message is not getting through. Dashboard signals sometimes indicate what is wrong, but electronic diagnostic service equipment knows much more. The two computers "talk to" each other as they run through a series of questions and responses to see how everything is running. That is how an automotive technician knows which parts or modules need changing.

New cars can run like a dream. But when something goes wrong, every driver needs an automotive technician who understands the car computer.

Teams of technicians include specialists in emissions, timing, mechanical and electrical systems, and pollution control. As they check out all parts, a report of problems and "OK" messages is printed out. These probably include transmission and drive line, automatic transmission, automotive suspension system, engine drive train group, variable suspension systems, wheels and brakes, timing, antilock brake systems, engine, and fuel injection systems. The test computer gives information back and forth as it goes through its checkpoints.

Other Services

Servicing is important for computer equipment of many kinds, such as microwave ovens, telephones, synthesizers, keyboards, sound systems, stereo, theater sound, 64" TV sets, or surround sound (five speakers connected by computer systems to surround listeners with music).

Look around. You'll see electronics everywhere. Even though computerized equipment is highly reliable, it all needs service on occasion. A good service technician becomes a hero to nervous or frantic equipment users.

Questions to Ask Yourself

There are many jobs in technological service. Any one of them may be perfect for you. 1) Do you like the idea of coming to the electrical rescue? 2) Are you willing to be on call for long hours? 3) Which field of service appeals most to you?

CHAPTER 6

COMMUNICATIONS

Check the credits on any movie or TV show you enjoy. The names are real people who enjoy using technology to create special effects and getting paid very well for doing it.

Technology in the entertainment field depends on computers and the skills of the people who operate them. Colors dissolve and reassemble. People age before your eyes. Action races across starry skies. Raisins dance and sing. A tiny covered wagon travels across a kitchen floor. Santa's sleigh appears on the roof. A monster bursts out of the ocean.

Creations like these from the hands and minds of technicians pushing the limits of computerized action emerge larger than life, large enough to fill the biggest theater screens. Someday they could be yours if you prepare yourself.

Millions of people watch television every day. Cable television and the information superhighway will add even more stations. Home computers will interact with TV. More technology jobs will open up. Some of them will be in your hometown.

Technical workers are needed to operate cameras, sound, lighting, and special effects. More jobs are in radio, theaters, sports arenas, churches, business, and home communications.

This is a technology job field that sizzles, but it takes long hours and routine work to make the sizzle happen.

Take music, for instance. Technology is changing music in dramatic ways. With computers and synthesizers, one person can perform all the sounds of a full orchestra, a rain forest, electrical storms, a marching band, crashing ocean waves, and more.

It looks easy, but it takes a great deal of skill and knowledge. Music is fun to listen to. It is fun plus pride mixed with hard work to create and control the sounds with chips, switches, wires, buttons, and mikes.

This kind of technology requires teamwork. Just as a quarterback can't score without the whole team, performers depend on technical teamwork to carry the ball.

Without them, lights and special effects wouldn't flash. Special sounds wouldn't be heard.

Studios require banks of electronic sound, editing, and controlling equipment, generally

Technology plays a large part in creating the music you listen to.

located behind glass, and skilled technicians doing hands-on jobs.

Technicians record and adjust sound. Others record images. They put them together, change tone and volume, and adjust speed. They mix sounds and block out what is unwanted. Technicians may become specialists in audio or video or both. They must understand the whole process to be a good team member.

A new kind of adventure is emerging: *virtual reality*, sometimes described as high-tech playtime. Virtual reality is more like being in the action than watching it. People are betting their future on virtual reality centers and are making

41

big plans to develop a new entertainment field through better technology.

Telecommunications technicians also work with many new ways of transmitting voices and information data across long distances. They link up telephones, computers, fax machines, and teletype machines. They also receive and send messages and signals faster and clearer by satellites, communication lines, laser beams, microwave transmissions, radar, and fiber optics cables. All open up new technical support jobs for the first time.

Interactive multimedia is a new industry that combines pictures, sound, graphics, and text. We see this in video games, educational software, and big-business presentations including those at Epcot Center and Disneyworld. New jobs in multimedia may not even have names yet, but high-tech skills in all phases of communication electronics are needed to present stories and messages in more interesting ways. The next generation of workers will grow with this industry.

The hottest of the new ideas is the coming information superhighway. This superhighway will be the total of all computer networks in the world. It will be like a great city with thousands of rooms full of people talking about everything from scientific discoveries to the best place to buy an out-of-print book. You can ask a question and find whatever information you need. School will be different when 26,000 schools are hooked up to the electronic superhighway.

All this means new communication jobs. Most job openings will be in smaller studios. Competition will be fierce for the best jobs in the biggest studios.

Somewhere in the launching of this electronic superhighway may be the job of a lifetime for you—if you are prepared and ready.

Questions to Ask Yourself

The fast-paced world of communications is a perfect place for technology to grow. Consider if a job in this area interests you. 1) Do you work well with other people? 2) Would you enjoy being in some part of the entertainment industry? 3) Are you willing to work long hours to see your work come together?

CHAPTER 7

PREPARING YOURSELF

What does it take to prepare for technology jobs?

Nothing will take the place of what you know and what you have experienced.

What you learn and what you do during your school years is your basic education. Use your high school years to choose mathematics, science, and communication classes. Developing your own brain power will pay off when it counts.

High school students can take advantage of JETS (Junior Engineering Technical Society) programs. JETS volunteers bring films, contests, field trips, clubs, and National Aptitude Search exams to school to show you what manufacturing technology jobs are like. They help you to get hands-on experience, show you how math and science apply to real-life situations, and answer questions.

Technology careers require education beyond high school. A two-year community college or a qualified trade school is a good choice. Be sure to

Experience and a good education will prepare you for a job in technology.

consult the financial aid office for scholarship help, grants, or loans. Apprenticeships also can help you to master the skills you are seeking.

Many technician jobs do not require a four-year college degree, but employers often prefer college graduates for the highest-paying jobs.

College and university courses are expensive, but it is possible to earn a degree while working. Joan and Pablo took community college courses a few at a time while working. Joan's employer paid some of her college costs. Pablo could see for himself what engineers do. He began to realize that he could do it too if he studied harder and a few years longer.

Computer technology in many fields demands employees who are well trained and comfortable with computers. A job must be earned. Getting ready means studying and building your skills at every level. Math, science, physics, computers, and language skills are the best ticket to the future. When you are really ready, employers will want you.

Starting early to use some of that information helps. Hobbies, experiments, and games develop your interests and computer skills and keep your mind sharp.

Many technology workers, especially computer and electronics technicians, start as assemblers, working after or between community college classes. This is hands-on work, and you only need to be familiar with one specific assembly function. The pay is low, but it gives you a chance to begin, to see inside electronics, to prove that you are a steady worker, and to pay for more skills classes. Assembly work can lead to testing, inspection, quality control, or technician jobs.

Companies that promote from within post job notices to give employees first chance. This is known as climbing the company ladder.

Military service is another good way to prepare yourself for technology jobs. The Air Force operates one of the largest first-rate technical training programs in the world. They offer free, no-obligation testing while you are still in high school.

Enlisting is not easy. You have to qualify by hitting the books in high school. Mathematics and science are most important. They also want English and literature with a GPA (grade point average) of 3.75 or 4.0. You compete with other students to be accepted. If you make it, they will also pay you, feed you, and provide housing while you're training. Staying in the military can lead to a good career.

Moving from one technician job to another is good for several reasons. Each skill learned is helpful in understanding technology. Some jobs relate to each other, and understanding makes for better teamwork. You can discover which type of work you like best and do best. Moving up to more complex jobs a step at a time brings bigger paychecks.

Reading electronics and other technology magazines helps. Trade journals on file are in the business section of your library. In them you can read about new applications of technology to spot trends for future jobs.

Volunteering is a good strategy. Create a computer data base for a local club, athletic team, or church group. Work backstage with school plays or local theater lighting and sound production crews. Hang around a neighbor who repairs technical equipment, helping and asking questions.

Serving in the military is another good way to learn necessary skills.

Hands-on activities such as building scale models or tinkering with fix-it projects help build skills. Give adults a chance to encourage you; many like to see young people get ahead. They may be good references and help you in your job search.

Questions to Ask Yourself

Preparation is essential for anyone hoping to have a good career in technology. 1) What classes can you take to help you prepare for your career? 2) Would you consider enlisting in the military? 3) Whom can you talk to about careers in

technology?

CHAPTER 8

FINDING A JOB

Finding the right job may not be easy. Investing time and energy in your job search will strengthen your chances of success.

Suppose you were the boss and received 40 job applications. Many were messy, a few were fairly easy to read, and three were clear, clean, and answered questions as if the writer really wanted the job. Which would you choose?

It's easy to think, "Who cares?" The truth is, the employer cares. He or she issues paychecks to workers and cares very much about finding good workers. He or she looks for the person who wants to do a good job. Getting the right kind of attention from the start really helps.

All job applications ask basic questions like name, address, phone number, social security number, education, and experience. Think ahead about your answers. Take time to write the information clearly and legibly.

49

Résumés

Writing résumés is a key to your future. A résumé is a concise biography of yourself—your education, your experience, and the kind of work you are looking for. Take time to do the best you can. Since you are seeking a job in a high-tech world, it makes sense to prepare your letters and résumé on a computer.

Make sure they look businesslike. Get help from a good typist or résumé specialist if necessary, but what your letter and résumé *say* is most important of all.

Job counselors can help you discover your strengths, and that information should be included. The more strengths, the better. You know yourself and are responsible for the accuracy. What you say must be true.

Give serious thought to your career choice and plans. Discover what you really want to do, your goals. "This is what I can do" statements are the best way of getting noticed. Employers look for people who will do the job that needs doing. If their job is what you like to do, say so. If not, try somewhere else.

Résumés are typed on white paper. List the jobs you've already held, volunteer work, your skills, school subjects, honors, clubs, or group service experiences such as Scouts. Be sure to list part-time work and experience. Include dates of graduation, jobs, or other experience. Books in the library can help you organize your résumé.

Résumés help employers find the right person for the job. The purpose of the résumé is to open the door.

Interviews

This is where you'll meet face to face with people who have the power to hire you. One question that you will most likely hear is, "Any high-tech background?"

If you are not prepared to say, "Yes. This is what I can do—", you will hear a quick "Sorry" and be on your way out.

You can prevent that. If you really want a good job, the money, and personal satisfaction, you'll get ready and then go after the job you want.

In an interview, your body speaks for you also. First impressions count. Dress neatly. Looking alert, making eye contact, and nodding your head show that you understand what is being said. Natural, confident smiles (not grins) are a powerful tool. Remember good posture, and have neat, clean fingernails. Never drink during the 24 hours before an interview. Many companies require drug testing as a condition of employment. Do not fidget with your fingers or your feet.

Attitude is the key word in a job interview. Both you and the employer are trying to see if you match. Questions and answers are the only

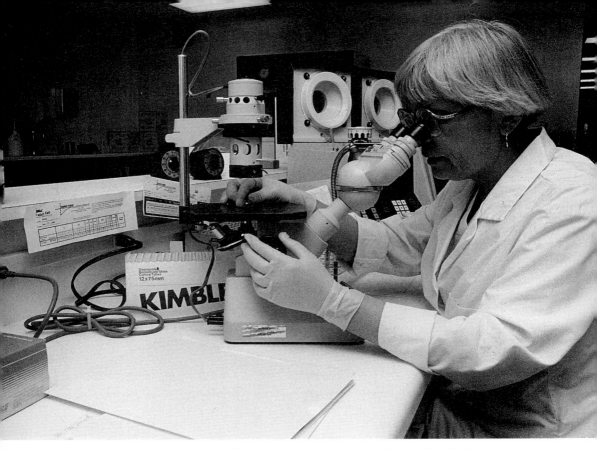

If you like the medical field, perhaps a job as a medical technologist is appealing. This technologist is reading blood test results.

way to do this. When both of you are satisfied, it's a match.

There is good news for young job-hunters. Since you started with computers when you were very young, you feel much more at home with them than do older workers. It is possible to increase that advantage even more. Give yourself extra know-how by using a computer at home. Earn one by doing chores and small jobs. Invest in yourself and your future by saving to buy your **52** own computer—before you even think of your own car. Get a car later when you're earning more.

Research which companies you'd like to work for. Which ones seem interesting? Find a way to meet people who do that work already. Places where you might meet someone who could help are the Chamber of Commerce, your church, community college instructors, or stores that sell supplies to the businesses. Ask questions. Find someone who knows the answers.

Do a school report on your research. Not only will you be increasing your own knowledge, but you will also improve your grade, your opportunities, your skills, and your self-confidence.

Geography plays an important role in your job search. There are clusters of high-tech businesses in and near big cities. You may do better if you are willing to move.

Your job-finding skills should be always kept up-to-date. Whether you want to find a job or move up to better jobs, being ready to sell yourself is an advantage.

Questions to Ask Yourself

Getting a job in technology isn't easy. You have to be prepared for it. 1) What would you put in your résumé? 2) How can you get experience now that will help you to get a good job later? 3) Are you willing to relocate for a better job?

CHAPTER 9

KNOWING YOURSELF

It is wise to know yourself on the way to choosing a career as well as during the many changes you will probably make during your working years. It may sound odd to say "know yourself," but many people do not. They just go along from day to day without thinking much about their strengths and what they would really like to be and to do. Life can be much better than that when you take time to know yourself.

For instance, ask yourself these questions and see what you discover:

- Are you dependable?
- Can you identify your strongest abilities and skills?
- Are you a self-starter? Or do you wait until someone else tells you what to do?
- Can you clearly explain what you do well and enjoy doing?

- How is your energy level? Are you strong enough to work more than eight hours a day?
- Are you honest?
- Do you care enough to look neat and clean?
- Can you manage your time and get things done when they are needed?
- Do you understand directions and know how to follow them?
- Do you usually do more than asked if you see something that needs fixing?
- Are you good at making decisions?
- Can you imagine better ways of doing things or creating new products?
- Do you organize and clean up after yourself when work projects get messy?
- If a job requires shift work, can you adjust to sleeping day or night?
- Can you get along with people of varying personalities?

These questions will give you an idea of what an employer is likely to expect of you. They will also give you an idea of what you can bring to a job in addition to technical skills. If your employer is happy with you and you are happy with your job, that is a win-win situation.

- What do you really like to do?
- What do you do best?

These last two questions are very important when preparing for a career. You work your best when you find what you like to do and then find someone to pay you to do it.

When you like technology and are excited about it, you have many job choices. Find out about as many of them as you can.

Having enough money to live on is a basic need. Think about how much money you need now and also how much you will need in a few years, especially after marriage and children. Entry-level technician jobs usually start at $15,000 a year or more. The pay varies with type of work, difficulty, responsibility, and geographical location. Cost of living such as rent and food varies too. Compare salaries and cost of living to see if they're a good match. Some technology occupations pay up to $40,000 or more, depending on the employee's experience. Some jobs lead to better growth opportunities later. Know your needs along the way and ask questions.

While money is important, it is not everything. You will be working for many years. If your work is interesting to you, you will feel more satisfied.

Some jobs can extend into your off-hours. You may like that—or you may not. Either way, make some of your choices on what is important at home as well as on the job.

Health factors may affect your job choice and performance. In some jobs, physical or mental

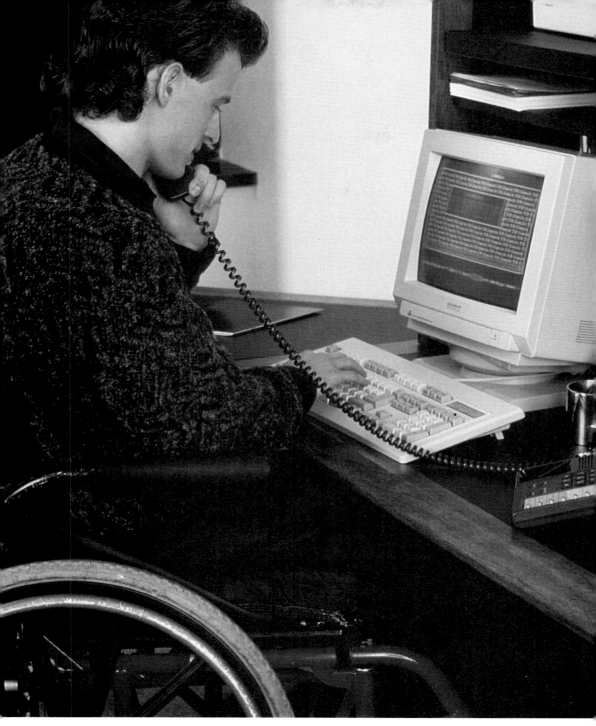

Computers allow people with disabilities to enter fields that might otherwise have been closed to them.

influences such as strain, stress, or long hours can affect your work and your body. Job hazards such as fumes, long periods of standing, back injuries, or eye fatigue may or may not cause physical problems. Be alert to what your body may be telling you to understand whether you are a match for a particular job.

For women, men, minorities, and those who have some type of disability, computer careers prove themselves over and over again as a great equalizer. Elan, for instance, proved she could run the college computer lab despite being unable to hear. In her first job with a major corporation, she initiated extra work and moved steadily up to better jobs. Her hearing dog went to work with her, and both were valued by the employer.

Computers can create more equal opportunities for all.

Questions to Ask Yourself

A career in technology may sound exciting to you, but think carefully about it before deciding to pursue it. 1) Will you become bored with a job in technology? Are you interested in the subject enough to truly enjoy your job? 2) Do you have the necessary skills and personality traits for your chosen specialization? 3) Are you physically capable of doing the job you want?

GLOSSARY

apprenticeship Program in which persons learn a trade by working under the supervision of a skilled worker. Apprentices often receive classroom instruction in addition to their supervised practical experience.

computer Machine that can follow instructions through the use of electrical signals to use and change information.

environmental Concerning the air, water, earth, plants, and all living creatures.

information superhighway Network of electronic mail, information sources, and telecommunications systems.

integrated circuit See *microprocessor chip*

mainframe computer Extremely large computer. Many workers can use the same mainframe by use of individual terminals at workstations.

microprocessor chip The internal chip that is the heart of a computer.

monitor Display screen used to view temporary images being processed by a computer or other high-tech testing device.

paraprofessional Specialized technologist who works in a professional setting.

résumé Listing of what you've learned, what you've done, and what you're equipped to do in your next job.

robots Devices that can grasp, move, and turn objects to assemble, install, or paint a variety of parts of manufactured items.

schematic Drawing showing how to make an electronic system.

service To install, repair, or improve functioning of equipment.

technician Worker with specialized practical training in a scientific or mechanical subject who works under the supervision of engineers, scientists, or other professionals.

technologist Worker in a scientific or mechanical field with one to two years more training than a technician.

telecommunication Electronic transfer of information from one place to another.

APPENDIX

Helpful Organizations

Automotive Service
  Association
  1901 Airport Freeway
  Bedford, TX 76095

American Hospital Association
  840 North Lake Shore
  Drive
  Chicago, IL 60611

Educational and Training
  Division
Robotics International
American Society of
  Manufacturing Engineers
  3 West 36th Street
  New York, NY 10001

American Society of Certified
  Engineering Technicians
  P.O. Box 371474
  El Paso, TX 79937

Association for the
  Advancement of Medical
  Instrumentation
  3330 Washington
  Boulevard
  Arlington, VA 22201

Junior Engineering Technical
  Society (JETS)
  1428 King Street
  Alexandria, VA 22314

National Action Council for
  Minorities in Engineering
  3 West 35th Street
  New York, Ny 10001

National Association of Trade
  and Technical Schools
  2251 Wisconsin Avenue
  NW
  Washington, DC 20007

Society of Broadcast Engineers
Information Office
  7002 Graham Road
  Indianapolis, IN 46220

Society for Technical
  Communications
  815 15th Street NW
  Washington, DC 20005

61

FOR FURTHER READING

Bone, Jan. *Opportunities in CAD/CAM Careers.* Chicago: VGM Career Horizons, 1994.

Eberts, Marjorie, and Gisler, Margaret. *Careers for Computer Buffs and Other Technological Types.* Chicago: VGM Career Horizons, 1994.

Exploring Careers. Indianapolis: JIST Works, Inc., 1990.

Gould, Jay, and Lusano, Wayne A. *Opportunities in Technical Writing and Communication Careers.* Chicago: GM Career Horizons, 1994.

Lytle, Elizabeth Stewart. *Careers as an Electrician.* New York: Rosen Publishing Group, 1993.

Morgan, Bradley J. and Palmisano, Joseph M. *Computing and Software Design Career Directory.* Detroit: Visible Ink Press, 1993.

Southworth, Scott. *Exploring High-Tech Careers,* rev. ed. New York: Rosen Publishing Group, 1993.

Stair, Lila B. *Careers in Computers.* Chicago, VGM Career Horizons, 1991.

Williams, Linda. *Careers Without College: Computers.* Princeton: Peterson's Guides Inc., 1992.

INDEX

ABOUT THE AUTHOR

Jean W. Spencer is the Public Relations/Information Officer of Oxnard College, a life member of the Ventura County Public Information Communicators Association (PICA), a member of the California Writers Club, the Society of Children's Book Writers, and the Conejo-Ventura Macintosh Users Group. She has served on the Commission on Public Information of the California Association of Community Colleges.

Ms. Spencer has been published regularly in more than a dozen newspapers on both coasts and more than two dozen national magazines and books. She is a computer enthusiast and a computer literacy advocate.

COVER PHOTO: © Barros and Barros/Image Bank

PHOTO CREDITS: p. 2 © Loren Santow/Impact Bank; p. 6 © Jay Freis/Image Bank; p. 9 © Harvey Finkle/Impact Visuals; pp. 12, 30, 52 © Martha Tabor/Impact Visuals; p. 14 © Tim Bieber/ Image Bank; p. 17 © Barros and Barros/Image Bank; p. 19 © Kay Chernush/Image Bank; p. 22 © Arthur d'Arazien/Image Bank; p. 24 © Terje Rakke/Image Bank; pp. 26, 31, 48 © Jim West/ Impact Visuals; p. 27 © Gary Gladstone/Image Bank; p. © Bill Varie/Image Bank; p. 41 © H. L. Delgado/Impact Visual; p. 45 © Evan Johnson/Impact; p. 57 © David W. Hamilton/Image Bank

PHOTO RESEARCH: Vera Ahmadzadeh

DESIGN: Kim Sonsky